OCT 25 1987

Withdrawn

J621.1 Siegel, Beatrice.
        The steam engine

$10.85

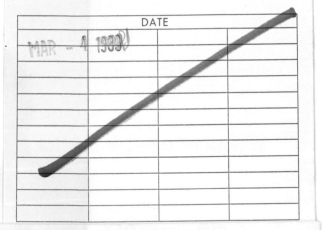

L 113

© THE BAKER & TAYLOR CO.

The Steam Engine

INVENTIONS THAT CHANGED OUR LIVES

The Steam Engine

Beatrice Siegel

Walker and Company
New York

Copyright © 1986 by Beatrice Siegel

All rights reserved. No part of this book may be reproduced or transmitted in any form or by any means, electric or mechanical, including photocopying, recording, or by any information storage and retrieval system, without permission in writing from the Publisher.

First published in the United States of America in 1986 by the Walker Publishing Company, Inc.

Published simultaneously in Canada by John Wiley & Sons Canada, Limited, Rexdale, Ontario.

Library of Congress Cataloging-in-Publication Data

Siegel, Beatrice.
   The steam engine.

   (Inventions that changed our lives)
   Includes index.
   Summary: A history of the steam engine, emphasizing that many people contributed to its invention and that it forever changed man's life for the easier.
   1. Steam-engines—History—Juvenile literature.
[1. Steam engines] I. Title. II. Series.
TJ467.S55 1986   621.1   86-5616
ISBN 0-8027-6655-2
ISBN 0-8027-6656-0 (lib. bdg.)

Printed in the United States of America

10 9 8 7 6 5 4 3 2 1

*To the memory
of my brother
Henry*

Contents

Preface

Chapter 1　MUSCLE, WATER, AND WIND　　1

Chapter 2　CAPTURING STEAM POWER　　11

Chapter 3　JAMES WATT　　21

Chapter 4　THE IRON HORSE　　29

Chapter 5　THE GOLDEN AGE OF STEAM POWER　　41

Index　　49

Preface

TODAY WE CAN fly in space, land on the moon, and sail huge ships over the oceans. To do these dazzling things—along with everything else we do—requires the use of energy in one form or another.

What happened in the days of our earliest ancestors? They lived long before the invention of a single machine or tool, and before animals were domesticated. Yet they survived. Their energy came from their own muscle power. Thus we call muscle power the first prime mover. It enabled our ancestors to build homes, hunt for food, and make clothing out of fur skins. Over the years people learned to extend their muscle power by shaping bone and wood into tools. They tamed animals and began the search for other sources of power. In time they harnessed water power and wind power.

In the never-ending search for energy sources, the invention of the steam engine changed the face of the earth. It helped launch the industrial world, making possible steamboats, locomotives, and large factories and farms.

The use of steam power is continuously updated through new technology. It remains a valuable energy medium even in today's world.

A steam locomotive coming around a bend.

The Steam Engine

Steamboats in a midnight race on the Mississippi.
Currier and Ives: Museum of the City of New York

1

Muscle, Water, and Wind

IT TOOK ONE HUNDRED THOUSAND men twenty years to build the great Egyptian pyramid Cheops. Considered one of the wonders of the world, Cheops is a monumental mass of six million tons of stone covering thirteen acres of land near Cairo.

Teams of slaves, organized into work groups, harnessed their muscle power to quarry, haul, and shape massive blocks of limestone and granite. Fortunately, by the time of the pyramids, workers had learned how to supplement their muscle power, using newly invented tools such as chisels, leveling instruments, stone and copper cutting implements, and pulleys. Above all, they had domesticated animals. In Egypt, animal power helped raise water from the Nile River to irrigate

Human muscle power built the Great Sphinx and pyramids of ancient Egypt, 2500 B.C. *Metropolitan Museum of Art*

farmland. Human muscle power was therefore reinforced by the muscle power of dogs, horses, donkeys, and oxen.

For tens of thousands of years, however, life rested primarily on the giant strength of human labor. Men, women, and children provided the basic muscle power that produced the roads, temples, palaces, and monuments of ancient kingdoms.

Though imaginative people searched for ways to lighten the burdens of everyday life, progress was slow in capturing different forms of energy. But the search was fruitful when people observed the power in the waterfall and in currents of rivers. Here was a natural resource that could be harnessed. Ancient Rome reported the first inven-

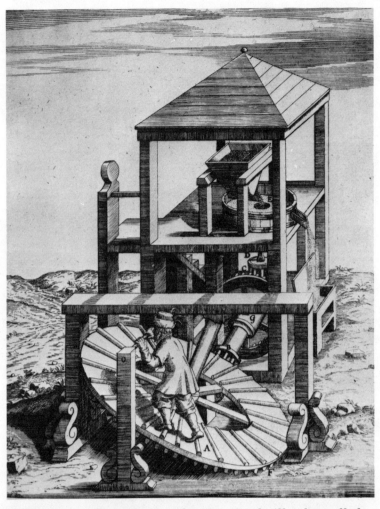

In the sixteenth century, a human treadmill, also called a footmill, provided power to grind grain. *Smithsonian Institution*

tion to use water as a source of power. Similar developments occurred in Egypt and China. The waterwheel was invented, and through many changes it developed into the one familiar today.

Dogs, oxen, horses, and other animals were harnessed to wheels or treadmills. *Smithsonian Institution*

Early waterwheels were made up of paddles, spokes, or buckets. When the wheel was placed in water, the flow of the current hit the paddles either at top or bottom. Where water entered at top, in the "overshot" wheel, the weight of the water added more force and power than was possible from mere current flow. Thus the waterwheel became the medium for converting the potential energy in river currents into mechanical energy.

The use of the waterwheel spread to other countries. It was put to work in all of Europe from

Water and wind power enabled people to be less dependent on human and animal muscle power. This waterwheel powers the machinery in a New England textile mill. *Smithsonian Institution*

the sixteenth to the eighteenth centuries. Water powered the machinery that ground grain, crushed stone, or raised water from deep wells.

Water power was also the foundation of industrial development in the United States. The power of fast running rivers was everywhere. In northern Maine, the combination of rivers and forest land made possible the lumber industry. Waterwheels built into rivers turned the machinery in sawmills that produced lumber for homes, ship building, and for export. Trade expanded when shipbuilding itself became an independent industry.

In New England, waterwheels powered textile mills. The production of vast supplies of fabric helped the clothing industry and led to the invention of the sewing machine.

At about the same time that water power was in use, another source of natural energy was harnessed—the power of wind. Simple craft, rigged with makeshift sails, navigated rivers in ancient times. But wind was unpredictable, and sailing ships in later years used a combination of wind and human muscle power. Teams of rowers or slaves, tied down in galleys, used their muscle power to move ships along when the wind died down.

On land, windmills became a major source of energy, especially in western Europe. They operated on principles similar to the waterwheel.

Wind power, another prime mover. *Smithsonian Institution*

Evenly spaced sails or blades were attached to a central axle and rotated with currents of air. The axle was attached through a system of gears to machines that ground corn, turned saws, or regulated water supply. Though the origin of the windmill was in seventh century Persia, by the twelfth century windmills were popular structures towering over the flat countryside of Holland. They regulated the canals that overflowed the low-lying land. Their use spread to England, France, and Germany. When the Dutch settled in New York in the early seventeenth century, they introduced windmills into the seaport town.

Human and animal muscle power, water, and wind were the prime movers for the human race through most of history. By the seventeenth and eighteenth centuries, however, these sources of energy were slowly being phased out by dramatic transformations in technology and society.

Surprising breakthroughs in mechanical inventions and craftsmanship sparked the period. They created social upheavals in which industrialists tried to increase their profits while workers fought for higher standards of living. At the same time the air crackled with debates in academic circles over different concepts in engineering and chemistry. New metals were being discovered as well as fossil fuels such as coal.

During the upheaval, determined inventors sat in attics and workshops, driving themselves to create original mechanical devices. They wanted to increase industrial productivity and speed up transportation and communication.

Uppermost in this period of scientific search were efforts to capture new sources of energy—energy that would come from a machine rather than from nature. Looming large was the advent of steam power.

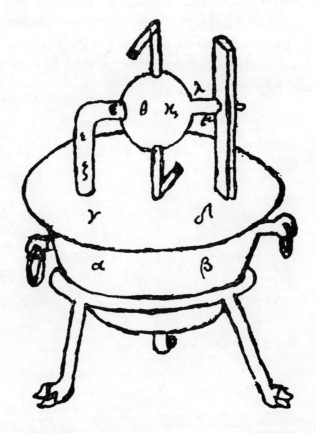

Hero's *aeolipile*. Steam, produced in the boiler, is led via pipes to the metal sphere which could turn on its axle.
The Science Museum, London, England

2

Capturing Steam Power

THE VISIBLE QUALITIES of steam have always been a source of fascination. We can see steam's characteristics in a simple form when we boil a kettle of water and vapor or steam escapes from the spout of the kettle. We can feel the heat and force of the steam if we accidentally put our hands near the spout. In more dramatic form, the steam of a covered pot of boiling water will make the lid dance up and down.

What happens if you put a very tight lid on a pot of boiling water and the steam cannot escape? The pot can explode. But if the pot does not explode, what happens to the steam captured within the pot? How can you capture steam and make it a medium that could drive a machine? These and other questions tantalized scientists and inventors for hundreds of years.

On record with the first mechanism to harness steam is the inventor Hero, or Heron, who lived in the days of the Alexandrian Empire. In about 200 B.C., the Museum of Alexandria (as the university was then called) was the center for learned men who taught advanced theories of mathematics and geography.

Hero, a colorful figure at the time, was excited by the potential of steam power. To harness it, he devised a machine he called the *aeolipile*. Today it is called a reaction turbine.

In the aeolipile, Hero placed a large covered vessel partly filled with water over a wood-burning fire. Two hollow tubes led the steam from the boiling water to a metal sphere directly above it which could turn on its axle. Exhaust pipes led out of the sphere at diametrically opposite ends and had openings at right angles to the axle. When the water bubbled, the steam was fed into the sphere. It then flowed through the sphere and escaped through the pipes, at the same time turning the axle and rotating the sphere.

Hero had used steam to obtain a rotary movement! His aeolipile had captured energy outside of the natural resources, muscle power, wind, and water.

Unfortunately, engineering technology had not sufficiently progressed to make the machine practical, and Hero turned the aeolipile into a toy to amuse his friends. Relic though it is, the aeolipile nevertheless remains on record as an example of the search for mechanical energy in ancient days.

Hundreds of years went by before the period of the Renaissance sparked a spirit of scientific inquiry. Scholars, notably in Germany, Holland, and France, debated an endless stream of theories. Along with the dynamic spurt of scholarship came advances in engineering and machine technology. Stimulated by this hectic activity, the seventeenth century produced major breakthroughs in steam engineering. Efforts were focused on the practical problem of raising water from the bottom of mine shafts.

The pioneer French inventor Denys Papin (1647–1714) spent his time studying the properties of steam. He was among the first to understand the nature of air—that air contained matter. He also grasped the significance of a vacuum, or a space completely devoid of any matter, even air. Along with this understanding came the revolutionary idea that if air contained matter, it could exert pressure, especially when it filled a vacuum. This insight, elementary today, raised scientific understanding to a new level in the seventeenth century.

Papin, as a professor in Marburg, Germany, continued his experiments with steam, air, and a vacuum. In 1690, he devised a miniature piston-in-cylinder engine. The piston was a rodlike piece that moved back and forth in a tubular-shaped enclosure called a cylinder. The back and forth movement was called reciprocating motion.

The Papin machine worked by creating steam in the space beneath the piston. The steam pushed up the piston. Papin then cooled the cylinder and condensed the steam, thus forming a vacuum. Atmospheric pressure on the other side of the piston pushed the piston back down the cylinder.

Papin was a visionary, a dreamer. He foresaw the day when steam pressure would operate carriages, ships, and factories. He dreamed steam would do the work and free people from backbreaking labor. Between Papin's visions and his capacity to make a workable machine lay years of failure. Technology still trailed scientific theory, and Papin never made a larger piston-in-cylinder machine. Unfortunately, he died in poverty. As so often occurs in the history of inventions, the glory and financial rewards were bestowed on those who came after him.

Thomas Savery (1650–1715) lived at the same time as Papin. He addressed himself to the problem of pumping water from the bottom of mines. In 1698, he built a machine called the "Miner's Friend." By creating a separate boiler for

Denys Papin
Smithsonian Institution

his engine, he added to the progress of the steam engine. But his machine suffered from similar weaknesses as Papin's device. Pipes often burst, or the boiler exploded, or there were not sufficient safety valves. But he achieved a considerable de-

Thomas Savery
Smithsonian Institution

gree of success when he applied steam pumps to power water fountains.

The first major breakthrough making the steam engine practical was the work of Thomas Newcomen (1663–1729). A resident of Dartmouth, England, Newcomen lived close to the old tin and coal mining areas which dated back to Roman times, the first century A.D. The ancient mines were deep, and water-driven pumps could no longer keep them dry. Newcomen, a merchant dealing in mine equipment, was familiar with the problems facing mine owners. He worried along with them that the mines would have to be shut down unless water could be pumped out of the shafts.

Aided by a skilled mechanic, John Calley, Newcomen spent ten years in concentrated work on a steam engine. Building on Papin's experience, he startled the world by producing a successful piston-in-cylinder machine. Known as the "atmospheric" steam engine, one was successfully installed in a mine shaft in Staffordshire, England, in 1712. Though the technology to build complicated parts of the engine was still backward, the engine successfully raised water from the bottom of deep mines.

In the Newcomen machine, a tap or faucet released steam into a cylinder fitted with a piston. The pressure of the steam forced the piston to move upward. When the piston rose, another tap released cold water into the cylinder to condense the steam directly. In the process, a vacuum was formed beneath the piston. Air or atmospheric pressure then rushed in on the other side of the piston and pushed the piston down. The cycle was repeated, heating and cooling the cylinder. Like Papin's machine, it was an atmospheric steam engine using the condensation of steam to create a vacuum into which atmospheric air pressure forced the piston. Though Newcomen also created steam in a separate boiler, he improved on it and cooled the steam by a water jet instead of cooling the cylinder itself.

The Newcomen machine did the practical job of raising water from mines as well as from deep wells. Considered the greatest single advance in the history of the steam engine, its success was confirmed by hundreds of orders.

Advanced as it was for its day, the Newcomen machine had intrinsic weaknesses which could not be remedied without further progress in mechanical engineering. The steam engine was made up of cylinders, pistons, valves, pipes, and joints. These frequently broke down because of inadequate manufacturing technology.

It was a broken-down Newcomen engine that inspired the young instrument maker, James Watt, to design a new machine.

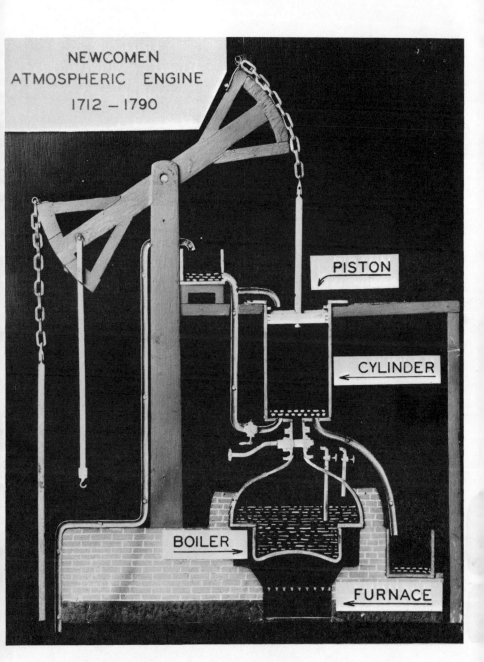

The Newcomen steam engine was called a "revolutionary giant." *Smithsonian Institution*

James Watt
Smithsonian Institution

3

James Watt

THOUGH HE IS generally credited with the invention of the steam engine, James Watt was only one of its inventors. He came at a time, however, when previous experimenters had already created workable but flawed machines. He improved on early efforts and made the steam engine completely practical. For this final achievement, Watt won acclaim and financial success.

Watt was born in 1736 in Greenoch, Scotland. At an early age he was apprenticed to a maker of scientific instruments. At age twenty-one he was appointed instrument maker at the University of Glasgow. While working in his shop in 1763, a Newcomen steam pumping engine was brought in for repairs.

Curious about the machine, Watt carefully investigated it to find out the cause for the breakdown. He also tried to find out why the machine ran out of steam so quickly.

Watt concluded that the loss of steam in the Newcomen machine resulted from the injection of cold water to make the steam condense. In the process, the cold water also cooled the cylinder itself. The alternate heating and cooling caused the loss of heat, slowed the process, and wasted power and fuel.

Watt applied his skills as craftsman and technician to work out a remedy for the Newcomen engine. In order to make the machine more efficient, Watt found himself redesigning it. In so doing he forever changed the nature of the steam engine.

Watt decided that the cylinder must always be kept hot, as hot as the steam. Yet, to condense steam, the vessel had to be cooled. To accomplish his purpose, Watt came up with a brilliant idea. In 1765, he invented a separate cylinder called a condenser. He placed the condenser next to the working cylinder. To make the working cylinder maintain its heat better, he fitted a steam jacket around it.

In 1769, Watt was granted a patent for "A New Method of Lessening the Consumption of Steam and Fuel in Fire Engines." This patent is considered one of the most important in the history of technology.

To improve production, Watt entered into partnership in 1774 with an enterprising businessman, Matthew Boulton. Under their firm's name, Boulton and Watt, the steam engine as prime mover became a major business enterprise. The first machine was installed in a coal mine in Tipton, England. The second worked the bellows of a blast furnace in Shropshire.

The machines aroused great public interest for they were more powerful than the Newcomen machine and could sustain their power without frequent breakdowns. English industrialists were particularly impressed that these machines used only a fraction of the coal of the Newcomen machines.

Orders for the steam engine flooded the Boulton and Watt office. The machines were in demand for tin and coal mines. And the widespread application of the steam-powered machines to other industries led to exploitation of fossil fuels such as coal.

To increase business further, the company promoted a policy of "leasing" machines, making them available to customers on a prescheduled fee. Sales soared and the Boulton-Watt steam engine continued its spectacular success.

Throughout his working life, James Watt continued to improve his machine. His inventive genius and his competitors standing in the wings with technologically updated engines spurred him on.

One of Watt's innovations was the use of a mechanical device called a governor. It regulated the speed of the engine and produced a steady motion even when the work load varied.

A major breakthrough came with the introduction of the rotary motion engine. A rod connected the piston to a crank, which is an axle or shaft bent at right angles. The turning of the crank converted the back and forth or reciprocating motion of the piston to a round-and-round motion. The rotary steam engine made it possible to power the wheels of industry, especially textile mills. Watt also attached a heavy wheel, called a flywheel, to the crankshaft to keep the rotary motion steady.

Among Watt's other inventions was the steam engine indicator to measure the engine's performance. And despite his fear of explosions, Watt increased productivity by increasing steam pressure.

In Watt's day, the use of the steam engine had progressed from simply pumping water out of deep mines shafts to become the prime mover for industry.

Both Boulton and Watt were humanists interested in the issues of the day. They founded the Lunar Society, a club for artists, writers, scientists, and businessmen, that met once a month when the moon was full. The full moon enabled members to find their way home along moonlit streets in the period before the use of streetlighting.

The nineteenth century became the era of the steam engine and dramatic industrial change. A spirit of hope spread among ordinary people. They daydreamed that mechanical energy would filter down to their daily lives and help them escape the burdens of work and oppression. Perhaps they would one day have the leisure to get an education, to read, and to travel.

The Boulton-Watt firm had produced five hundred steam engines by the time the patent expired in 1800. Watt, at his death at age eighty-three in 1819, was hailed as both an industrialist and a humanist for the benefits resulting from his invention.

The early steam engines, which were bulky, slow, and cumbersome, gave way to further improvements. Changes progressed chiefly along three main lines: higher steam pressure, improved utilization of steam expansion, and better mechanical design.

The high-pressure steam engine was the brainchild of a brilliant Cornishman, Richard Trevithick (1771–1833). Building on Watt's work, and less fearful than Watt of explosions, Trevithick devised higher pressure for the steam engine. He thus increased its capacity and speed, leading the way to small fast engines. He also dispensed with reliance on the vacuum. Instead,

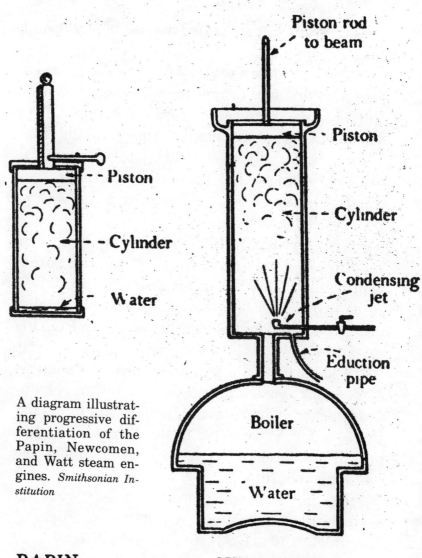

A diagram illustrating progressive differentiation of the Papin, Newcomen, and Watt steam engines. *Smithsonian Institution*

PAPIN   NEWCOMEN

Trevithick connected the piston rod directly to a pump or wheel and paved the way for applying steam power to transportation. In 1801, his engine carried the first passengers by steam power

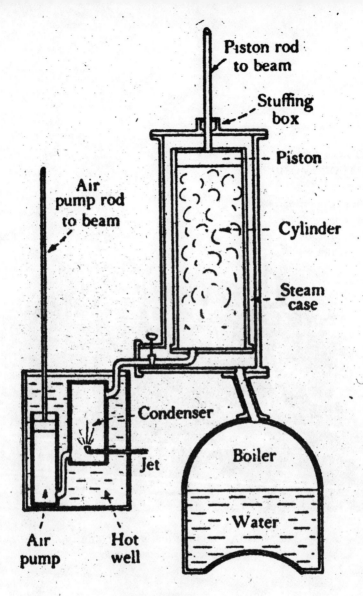

WATT

in London. In 1804, he built a steam locomotive that was used on a railway track in Wales. These inventions earned for Trevithick the well-deserved title, "Father of the Locomotive."

**THE WORKING PARTS OF
A STEAM LOCOMOTIVE**
1. Boiler
2. Cylinder containing piston
3. Piston rod
4. Connecting rods
5. Driving wheels

The working parts of a steam locomotive
Chicago and Northwestern Railway

4

The Iron Horse

THE UNITED STATES was only a cluster of British colonies along the Atlantic coast during the years that Papin, Savery, and Newcomen were working on the steam engine in Europe. The colonists were not thinking of inventions but of problems of surviving in the wilderness. By 1769, when James Watt invented the condenser and made steam power practical, a few colonists had started to talk about freedom from Great Britain. And while Europeans forged ahead with revolutionary inventions, the thirteen colonies fought a Revolutionary War. Their Declaration of Independence in 1776 came at a decisive time in the development of the steam engine. Steam power would be the means by which the emerging nation would transform itself into an industrial giant.

Independence from Great Britain brought an outburst of creative energy. Inventive minds and commercial and political forces set in motion dreams of greatness. One hundred years after independence, through war, purchase of land, and dispossession of the Indians, a huge slice of North America was incorporated into the United States. Designers of the American dream saw the stars and stripes fly over the country from the Atlantic to the Pacific oceans.

To open new frontiers, pioneer settlers at first traveled by stagecoach and wagon. They struggled through mountain passes, rivers, and prairies.

Meanwhile, the steam engine was creating miracles in transportation. Inventors abroad and in the United States were at work in machine shops ironing out details for mechanical transportation on land and sea.

Scotland was the site of the first practical steamboat launched on one of its rivers in 1802. Its success led to immediate further work on self-propelled vessels.

At the same time, Richard Trevithick, working on his high-pressure steam engine, had built a steam locomotive in 1804.

The scene then shifted from Europe to the United States where talented machinists were trying to catch up to their colleagues abroad.

Oliver Evans (1755–1819) loomed on the local scene in Newport, Delaware, where he was a millwright and engineer. He was involved in steam production the same year that Trevithick was creating the high-pressure engine. Evans, despite the backward condition of United States technology, also built a high-pressure steam engine. He placed it in a flat-bottomed dredging barge and used its power to move the barge from his works to the riverside, a distance of one and a half miles. He thus created a steam-powered moving vehicle. His engines were also driving sugarcane mills and grist mills. Despite some success, Evans's work was not fully appreciated, and he died a poor and disappointed man.

Like Oliver Evans, John Fitch (1743–1798), a Connecticut clockmaker turned inventor, was hampered by a lack of funds and a disbelieving public. He was a great artisan, however, and he dreamed of steamboat river navigation. After years of hard work and many failures, he laid the groundwork for a self-propelled boat. His awkward-looking craft was moved by a series of twelve paddles and a chain. He also tried paddles at the rear of the boat. Though he launched and developed passenger service on the Delaware River, he considered his life a failure and ended it in suicide.

Success in launching a steamboat capped the life of Robert Fulton (1765–1815), a Pennsylvania artist. Fulton had been studying portraiture in England and France when he unexpectedly changed course and became deeply involved in maritime engineering. At first, he gave his time to canal construction. After that he turned to submarine warfare and built an experimental submarine called the *Nautilus*.

When he lived in France, Fulton became friendly with the United States minister to Paris, Robert R. Livingston. A man of great influence and prestige, Livingston was involved in steamboat engineering in the United States and had gained a monopoly for steam navigation on the Hudson River.

In 1806, after twenty years abroad, Fulton returned to the United States to settle. He brought with him a Boulton-Watt steam engine, and he had Livingston's full support and influence.

Like many inventors, Fulton had to overcome public ridicule. In 1807, while the Boulton-Watt engine was being fitted into a clumsy-looking craft, spectators dubbed it "Fulton's Folly." And when August 17 arrived, and the ship, the *Clermont*, was ready to steam up the Hudson, onlookers on both sides of the river hooted in scorn, expecting to see the vessel explode. But the *Clermont*, smoke pouring from its funnel, calmly chugged up river covering one hundred and fifty miles in thirty-two hours. Fulton was acclaimed and reaped the heady rewards of public success.

The partnership of Fulton and Livingston dominated all Hudson River steamboat travel. Their ships maintained a regular passenger service between New York and Albany. Commercial

The *Clermont* off the Battery. *Museum of the City of New York: lithograph by J. H. Sherman*

companies developed similar monopolies on the Mississippi, Ohio, Delaware, and other great rivers. In 1819, the ship *Savannah*, sailing partly under steam power, made the first Atlantic crossing by a steamship.

The highpoint of steam power, however, was symbolized by the steam locomotive. The Indians called it the "iron horse." Indeed, it looked like an iron monster breathing flame and trailing clouds of smoke and acrid fumes across an unspoiled landscape. Hurtling through tiny backwater hamlets, the locomotive hauled passengers and freight. Even when stationary, the giant wheels of the trains and tons of iron and steel symbolized power.

The steam locomotive started small, a fantasy in the minds of engineers and inventors in restless pursuit of a mechanical means of transportation.

How did James Watt's steam engine, that used steam to push a piston up and down, make a train move? The basic elements were the same. Fuel, at first wood and then coal, was used to heat water in a boiler where the water was converted to steam. The force of the steam was piped to cylinders where it pushed steel pistons back and forth. The pistons were connected to the wheels of the locomotive by means of rods. The pistons, moving back and forth, turned the wheels that made the locomotive move. When steam pressure increased, the pistons moved faster and the train picked up speed.

In England, Richard Trevithick worked on a steam carriage in 1804. To arouse public interest in his train, he called it "Catch Me Who Can" and ran his train on a circular track near Euston Road, London. People lined up to pay a fee to become passengers on the four-wheeled carriage pulled by Trevithick's steam locomotive.

Taking advantage of improved technology, another skilled English machinist, George Stephenson (1781–1848), built the "Rocket," a locomotive that ran a track in Wales and pulled six cars of coal and twenty-one coaches of passengers. It was a resounding success. By 1830, Stephenson's engines operated the first regular railway passenger service in the world.

United States inventors were keeping pace with Trevithick and his successors in England. Building on Oliver Evans's first steps with the self-propelled land dredge, engineers began work on the steam locomotive. They were especially skilled in reshaping the work of British inventors to suit the needs of the United States.

By 1826, John Stevens ran a circular railway in Hoboken, New Jersey, at twelve miles an hour. In 1830, the Baltimore and Ohio ran its train called the "Best Friend." That same year, Peter Cooper, owner of iron works in Baltimore, had already designed a small locomotive he dubbed "Tom Thumb." Confident of its speed, Cooper entered the train in a race with a horse-drawn car. To his dismay, his steam engine broke down and came to a stop. The horse won the race, but Cooper was undaunted and continued his drive to make the railroad a major industry.

From small beginnings, scattered short railway lines in the Northeast expanded to longer lines reaching farther into the West. In 1852, the first trains from the East whistled into Chicago. By 1860, a network of iron rails crossed every state from Maine to Mississippi. Twenty thousand miles of track were laid in one decade. To make way for them, forests were cleared, mountain ranges blasted, river beds shored up, and rivers spanned. Blowing a mournful whistle, trains

drive a golden spike into the last railway tie. The whole country celebrated the engineering feat that linked one end of the continent to the other.

On May 15, 1869, regular service began on the first transcontinental train in the United States. People began to travel on trains hauled by steam-powered locomotives, hardly aware of the mechanisms up front that made the trains move. They knew little of engineers, firemen, and stokers who fed coal to the fire that kept steam hissing in the giant boiler. They were interested in the new inexpensive means of travel, in speed and comfort. Gone were the days of the stagecoach, of the river boat and canals. Speed had become the keynote to progress.

Over the years, lavish improvements replaced the dirty, noisy, smoked-filled trains of early days. After the 1870s, pullman sleepers, dining cars, plush seats, and elegant service made first-class passenger travel a luxury. Railway accidents slowly declined through installation of safety devices.

By 1900, a network of railway tracks covered the country, making possible a billion tons of freight a year. Though the impact of the steam engine is symbolized in the railway locomotive, it brought about other changes. Steam power created an upheaval in the economic structure that changed the very nature of society.

"Catch Me Who Can," Richard Trevithick's early steam locomotive on a circular track near London—*Smithsonian Institution*

United States inventor John Stevens trying out his railway Hoboken, New Jersey, in 1826—*Smithsonian Institution*

rushed through the homelands of tribes of Indians, into remote towns and villages, changing the focus of everyday life for thousands. They brought mail, newspapers, and manufactured goods into small towns and carried farm produce to urban centers.

Steam was supreme after the Civil War. The locomotive reached its peak with the completion of the first transcontinental railway in 1869. Supported by huge government subsidies of money and land, the Central Pacific built rails eastward from the California line while the Union Pacific built westward. Ten thousand men, mostly nese and Irish immigrants, laid miles of each day. They used explosives to cut their through the Sierra Nevada mountains; they from bridges over canyons and hammered r lines into the great plains. In the desert the Promontory Range, Utah, the two met on May 10, 1869. Hundreds of wor well as government and railway ex guarded by soldiers of the Twenty-firs States Infantry Regiment, observed th event. They saw railway official Lelan

East Meets West at Golden Spike ceremony in 1869 at Promontory Point, Utah. *Golden Spike National Historic Site: photo by Andrew J. Russell*

Portable steam engines
Smithsonian Institution

5

The Golden Age of Steam Power

STEAM POWER HAD insatiable needs. It continuously required new machines and replacements and tons of steel, iron, and coal. To amass these products, the steam power industry became the single greatest employer of labor. It put to work tens of thousands in its own and subsidiary plants.

Steam power created influential industrial figures and also invaded everyday life. The railway was like a magic carpet fulfilling people's hopes for a different future. They felt free to move from place to place. Many ended their rural isolation and relocated in urban centers. Others traveled west to farm. They dug for coal in Pennsylvania and for metal deposits around the Great Lakes. In 1861, the textile industry alone had

The Peak of Steam Power. A steam locomotive

coming around a bend.   *The Long Island Railroad*

Power is conveyed by belt from a steam engine to this large Singer machine factory in 1853. *Harper's Weekly*

grown to seventeen hundred steam-powered mills employing sixty thousand workers.

Farms and factories put out calls for more workers. The need for labor encouraged the immigration of millions of foreign-born. Their voices gave new shape to the country.

A significant aspect of steam power was the invention of the portable steam engine. Again everything dates back to the Englishman Richard Trevithick. In 1800, he combined the engine and boiler into one unit and put the assemblage on wheels. It was hauled from place to place at first by a team of horses; then it was self-propelled.

By 1850, portable steam engines were a common sight. The engine's power was transmitted by power transmission belt or other gear to a piece of machinery, to a farm thresher, a deep well, lawn mower, stone crusher, or even to a small factory. At amusement centers, the portable steam engine was found alongside the carousel making it go round-and-round.

By the end of the nineteenth century, portable steam engines were produced in the thousands and exported. Underdeveloped countries in particular found them useful because of their low cost and their ability to burn a wide range of fuels, including waste straw. They drove the machinery in small sugarcane plants and sawmills.

"The steam engine," says author H. W. Dickinson, "was never so important in the world's econ-

A portable engine powers a threshing machine. *Courtesy J. I. Case Company, Racine, Wisconsin*

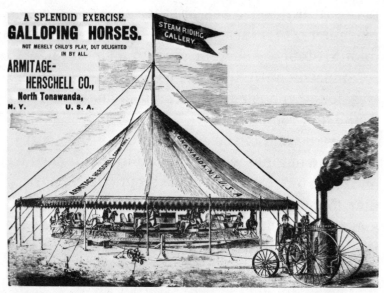

Here a portable steam engine transmits power to a merry-go-round. *The Painted Post Calliope Company*

omy as it is today."* He refers to sophisticated technological advances that produced the modern steam turbine—a large wheel or rotor made up of many sets of thin curved blades. Steam enters at one end and spins the bladed wheel which can be connected to an electric generator.

The steam turbine made its appearance in 1876. In the following years its power was transformed by generators into electric current that lit a million lights on highways and in homes. Today

---

*H. W. Dickinson, *A Short History of the Steam Engine*, Cambridge University Press, 1939.

steam turbines power electric generators in electric plants and drive huge naval vessels and steamship liners.

Steam power has not gone out of style but remains a vital energy medium in today's space age.

Index

Aeolipile, 10, 12–13
Albany, New York, 32
Alexandrian Empire, 12
Ancient Rome, 2

Baltimore, Maryland, 35
Boiler, 10, 14, 15, 18, 26, 27, 28, 34, 38, 44
Boulton, Matthew, 23
Boulton Watt firm, 23–27
Boulton and Watt steam engine, the, 23–29, 32
British colonies, the, 29

California, 36
Calley, John, 17
"Catch Me Who Can", the, 34, 37
Central Pacific Railroad, 36
Cheops, 1
China, 3
Civil War, the, 36
Clermont, the, 32, 33
Coal, 8, 23, 34
Condenser, 22, 27, 29

Cooper, Peter, 35
Cylinder, 14, 22, 26, 27, 28

Dartmouth, England, 16
Declaration of Independence, the, 29
Delaware River, 31, 33
Dickinson, H.W., 45
Dining cars, 38
Driving wheels, 28

Egypt, 1, 2, 3
Electric generators, 46, 47
Energy, 2, 4, 6, 8, 9, 19, 30, 47
Engineers, 38
England, 8, 31, 34
Europe, 4, 6
Euston Road, London, 34
Evans, Oliver, 30–31, 35

Firemen, 38
First transcontinental railway, the, 36–39
Fitch, John, 31
Flywheel, 24

Footmill, 3
Fossil fuels, 8, 23
France, 8, 13, 31, 32
Fulton, Robert, 31–32
"Fulton's Folly", 32

Galleys, 6
Glasgow, University of, 21
Golden Spike, the, 38, 39
Greenoch, Scotland, 21
Germany, 8, 13
Great Britain, 29, 30. *See also* England
Great Lakes, the, 41
Great Sphinx, the, 2

Hero, 10, 12–13
Heron, 12. *See also* Hero
Hoboken, N.J., 35, 37
Holland, 8, 13
Hudson River, 32

Indians (American), 30, 33, 36
"Iron Horse", the, 29, 33

Livingston, Robert R., 32
Locomotives, vi, vii; development of, 33–39; first regular passenger service, 35; fuel for, 34. *See also* Steam locomotive
Lunar Society, the, 24

Maine, 6, 35
Marburg, Germany, 14
Mechanical devices, 9
"Miner's Friend", the, 14
Mississippi River, ix, 33
Mississippi, State of, 35

Muscle power, vi, 1, 2, 5, 8; first domesticated animals, 1
Museum of Alexandria, 12

Nautilus, the, 32
Newcomen steam engine, the, 17–19, 21, 22, 23, 26
Newcomen, Thomas, 16–19, 29
New England, 5, 6
Newport, Delaware, 30
New York, City of, 32
New York, State of, 8
North America, 30

Ohio River, 33

Papin, Denys, 13–15, 17, 26, 29
Paris, France, 32
Pennsylvania, 31, 41
Piston-in-cylinder engine, 14, 17
Portable steam engine, 40, 44–45
Promontory Point, Utah, 39
Promontory Range, Utah, 37
Pullman sleepers, 38
Pyramids, the, 1

Reaction turbine, 12
Reciprocating motion, 14
Renaissance, the, 13
Revolutionary War, the, 29
"Rocket", the, 34
Rotary motion engine, 24

Savery, Thomas, 14–16, 29
Savannah, the, 33

50

Scotland, 30
Self-propelled boat, 31
A Short History of the Steam Engine (Dickinson), 46 n
Sierra Nevada Mountains, 37
Staffordshire, England, 17
Stanford, Leland, 37
Steam, 10, 11, 13, 14, 18, 22, 34, 36. *See also* Steam power
Steamboat, vi, ix; development of, 30–33; first practical, 30
Steamboat travel, 32; first Atlantic crossing, 33
Steam carriage, 34
Steam engine, vi, 15, 29, 30, 38; atmospheric steam engine, 17; development of, 11–21; high pressure steam engine, 25, 30, 31; improvements on, 24–25; in mills and factories, 44, 45; success of, 23–29. *See also* Steam power
Steam engine indicator, 24
Steam jacket, 22
Steam locomotive, 27, 28, 33–39, 42–43. *See also* Locomotive
Steam power, vi, 9, 12, 26, 29, 33, 38; golden age of, 41–47. *See also* Steam, Steam engine

Steam powered mills, 44
Steam pressure, 24, 34
Steam pumps, 16
Steamship liners, 47
Steam turbine, 46, 47
Stephenson, George, 34–35
Stevens, John, 35, 37
Stokers, 38
Submarine, 32

"Tom Thumb", the, 35
Treadmill, 3, 4
Trevithick, Richard, 25–27, 30–31, 34, 35, 36, 44
Twenty-first United States Infantry Regiment, 37

Union Pacific Railroad, 36
United States, the, 6, 29, 30, 31, 32, 35, 38

Wales, 34
Water power, vi, 3, 5, 6
Waterwheels, 3–6
Watt, James, 18, 20–27, 29, 34; death of, 25; develops new steam engine, 22–24; early life of, 21; other inventions of, 24; success of engine, 23–29. *See also* Steam Engine
Windmills, 6–8
Wind power, vi, 5, 6, 7